MOON
A First Look

PERCY LEED

Lerner Publications ◆ Minneapolis

Educator Toolbox

Reading books is a great way for kids to express what they're interested in. Before reading this title, ask the reader these questions:

> What do you think this book is about? Look at the cover for clues.
>
> What do you already know about the moon?
>
> What do you want to learn about the moon?

Let's Read Together

Encourage the reader to use the pictures to understand the text.

Point out when the reader successfully sounds out a word.

Praise the reader for recognizing sight words such as *the* and *we*.

TABLE OF CONTENTS

Moon

Look up in the
night sky.
What is that
bright glow?

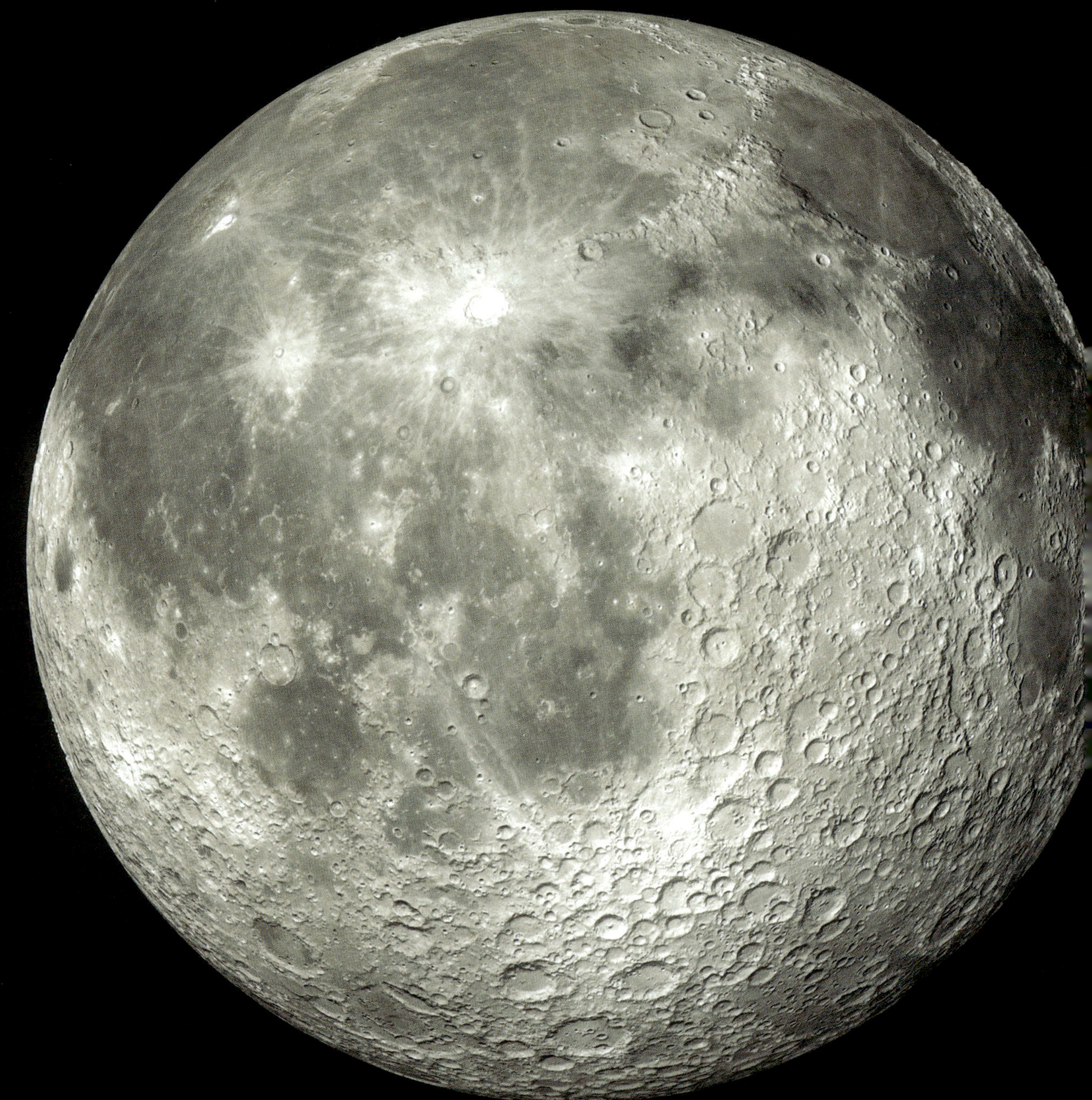

It is the moon.
The moon is Earth's
neighbor.

Earth has only one moon.

Other planets have
many moons.

The moon is
made of rock.
There are big
holes on the moon
called craters.

How do you think craters form?

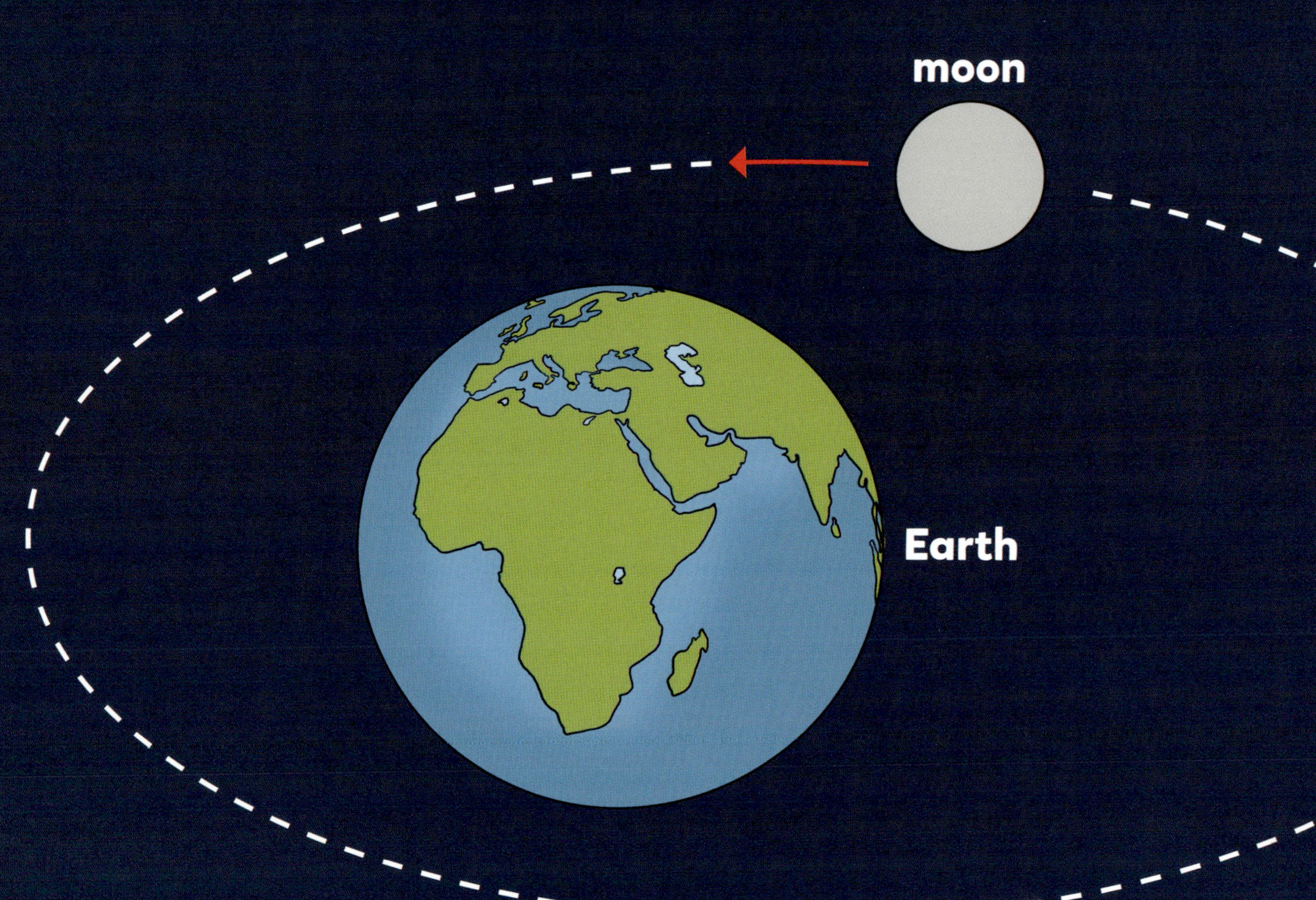

moon
Earth

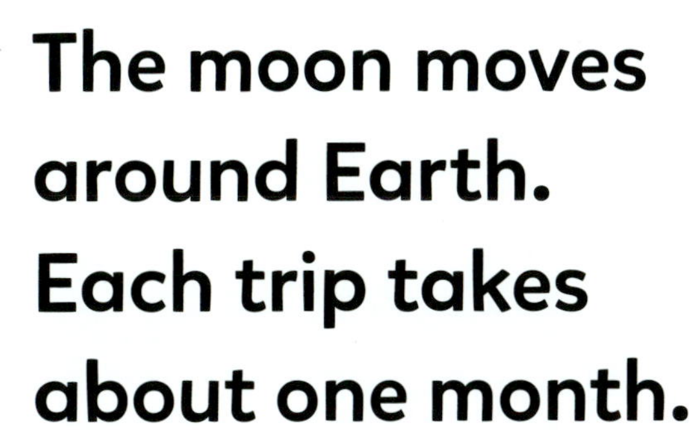

The moon moves
around Earth.
Each trip takes
about one month.

The sun makes
the moon glow.
Sometimes we see
the full moon.

Sometimes we see
half of the moon.

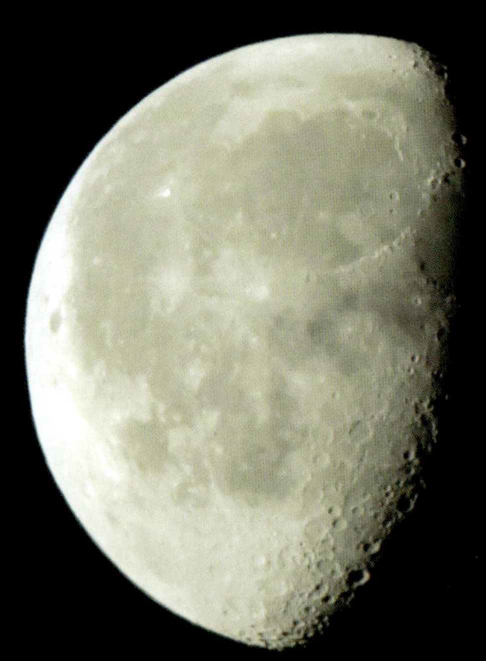

Sometimes we see only
a little of the moon.

In 1969, astronauts walked on the moon.

Do you know
how astronauts
travel to space?

Maybe you can go to
the moon one day!

You Connect!

What is something you like about the moon?

Have you ever seen a full moon?

Do you think you would like to visit the moon?

STEM Snapshot

Encourage students to think and ask questions like scientists. Ask the reader:

What is something you learned about the moon?

What is something you noticed about the moon?

What is something you still want to learn about the moon?

Photo Glossary

Learn More

Leed, Percy. *Earth: A First Look*. Minneapolis: Lerner Publications, 2023.

Lovitt, Chip. *My Little Golden Book about the First Moon Landing*. New York: Golden Books, 2019.

Rustad, Martha E. H. *Let's Notice Patterns in the Sky*. Minneapolis: Lerner Publications, 2022.

Index

Photo Acknowledgments

The images in this book are used with the permission of: © BoValentino/iStockphoto, pp. 14–15, 23 (full moon); © cyperc stock/Shutterstock Images, pp. 4–5; © Ed Williams/iStockphoto, p. 16; © Helen_Field/iStockphoto, pp. 10–11, 23 (crater); © Ioannis Pantzi/Shutterstock Images, pp. 8, 23 (Earth); © Just_Super/iStockphoto, pp. 6–7; © Marilyn Nieves/iStockphoto, p. 20; © NASA, pp. 18–19, 23 (astronaut); © NASA/JPL, p. 9; © ollirg/iStockphoto, p. 17; © renklerin kafasi/Shutterstock Images, pp. 12–13.

Cover Photo: © NASA.

Design Elements: © Mighty Media, Inc.

Lerner Publications Company
An imprint of Lerner Publishing Group, Inc.
241 First Avenue North
Minneapolis, MN 55401 USA

For reading levels and more information, look up this title at www.lernerbooks.com.

Main body text set in Mikado a Medium.
Typeface provided by Hannes von Doehren.

Library of Congress Cataloging-in-Publication Data

Names: Leed, Percy, 1968- author.
Title: Moon : a first look / Percy Leed.
Description: Minneapolis, MN : Lerner Publications, [2023] | Series: Read for a better world | Includes bibliographical references and index. | Audience: Ages 5–8 | Audience: Grades K–1 | Summary: "The moon is a familiar sight in the night sky. Easy-to-read text and fascinating photos give readers a first look at Earth's rocky neighbor"– Provided by publisher.
Identifiers: LCCN 2021051628 (print) | LCCN 2021051629 (ebook) | ISBN 9781728459226 (library binding) | ISBN 9781728464343 (paperback) | ISBN 9781728462196 (ebook)
Subjects: LCSH: Moon—Juvenile literature.
Classification: LCC QB582 .L44 2023 (print) | LCC QB582 (ebook) | DDC 523.3–dc23/eng/20211216

LC record available at https://lccn.loc.gov/2021051628
LC ebook record available at https://lccn.loc.gov/2021051629

Manufactured in the United States of America
1 – CG – 7/15/22